OFF-GRID SOL ⦿⦿54123

BUILD YOUR OWN

AFFORDABLE OFF-GRID

SOLAR SYSTEM

Author: John Slavio

TABLE OF CONTENTS

DISCLAIMER

ABOUT THE AUTHOR

John Slavio is a programmer who is passionate about the reach of the internet and the interaction of the internet with daily devices. He has automated several home devices to make them 'smart' and connected them to high-speed internet. His passions involve computer security, iOT, hardware programming, and blogging.

BENEFITS OF OFF-GRID SOLAR

Power Outage Possibility Decreases

As a person who has had to survive a hurricane or two, I can officially tell you that one of the best reasons to make your own Off the Grid solar power system is simply to have electricity when other people don't. Because you are the one that is monitoring and providing maintenance on your specific grid, this means that the likelihood of a random power outage is very unlikely, unless one of your solar panels is shattered by some unknown force.

In a giant power grid that connects wires that traverse tens to even hundreds of miles, practically anything can go wrong. A transformer could blow, a power line could be knocked down, one of the wires could snap and come loose etc. Any number of things can go wrong. When an electric company must find a problem, first they narrow the problem to specific sections of the grid, then they manually inspect all of the individual poles and lines in that section to find the problem. Locating and fixing the issue can take a long time depending on where it originates.

On the other hand, because your solar grid is typically located on your property, if you notice that you are starting to lose power you can easily detect it through monitoring and then find the source of the problem within a relatively small area. In your system, all you have to do is simply look at the wires that are connecting to your battery packs and see which one might be causing the problem. Then, if the wires all check out, you just look at the solar panels themselves to see if any of them have broken or if something happened to them inside of their cases. Finally, you can then check the charge controller to see if it got hit with a surge of power or you can check one of the many surge protectors that you lay around the house so that you don't fry your devices. In other words, whenever you do experience a power outage you know the exact reason why you are experiencing it, and you are busy fixing it rather than having to wait for somebody else to fix it.

Saving on Electricity ... Eventually

One of the benefits that the solar companies try to sell you on is the fact that you'll be saving money by switching to solar power, which is true but is also a really small aspect of actually building your solar grid. You see, your savings come from allowing it to pay for itself over time but the time it takes can vary. For instance, let's say that you consume somewhere around 1 kilowatt a day. That would mean that you would need a thousand-watt solar panel just to make sure that you have power, but you also need an additional one to compensate for the constant use of that power. The combination of these, if you don't build it yourself, usually means that you are spending anywhere from $400 to $1,000. Then, you have the charge controller so that you don't blow the system up and that's usually somewhere around $50 to $400. Next, you need batteries, which usually costs anywhere from $200 to $1,000 each. Finally, you have all the extra connections that are there for making sure that a surge doesn't touch your system and that's usually about $100. Therefore, if you add all of this at the maximum price to get an efficient system, then you will have spent somewhere around $3,000. If you only spend around $150 to $200 a month, then it will take you almost 2 to 3 years for this system to pay itself off. Now here's the kicker, most people do not realize that 1 kilowatt of power is absolutely laughable when it comes to talking about power consumption. Air conditioning alone takes more than one kilowatt of power which explains why most people under-build their system because they think that it simply shouldn't cost that much if they build it themselves. Most systems that are built by hand normally cost within the range of $5,000 to $15,000 depending on how much electricity you use. Therefore, if you do spend anywhere from $150 to $300, it will take almost five years to pay off one of the more expensive systems that you put on your house. Therefore, many people choose to build their solar grid incrementally.

Bring Electricity to No Electricity Areas

If you are a person who plans to be on the road all the time inside of an RV or you own a property that's nowhere near the actual electric grid, then you will find having a solar grid to be very beneficial.

When you are in these situations, you can use your solar grid to provide the electricity that you need.

A good example is a person who travels around in an RV and works from his or her home who may want to go to different locations because they don't need to be in any specific area to get internet access. The cost of paying for electricity in an RV park is all right but what if they don't want to park at an RV park? Sure, they could run their RV all night to provide the power that they need, but this would consume a considerable amount of gas either running the engine or the RV's backup generator. On the other hand, if they have a solar system that charges batteries that are housed somewhere in the RV, then they always have access to additional power that they can use in the middle of the night without having to use the extra gas in the RV or rely on a backup generator.

Now a person who owns a property that's nowhere near the actual electric grid might be better off supplying their own electricity. In most cases the cost of bringing electricity to remote regions is prohibitive. For example, in desert regions there are almost no poles that carry electricity beyond the major transmission lines between cities and even this is rather uncommon because of the distance. It's usually cheaper for electric companies to provide the electricity locally rather than carry it over lines from another area, which creates regions with no easy access to the grid.

Rising Electric Costs

As I already mentioned, it's not really about the end game savings that you get from purchasing your first solar system but rather how much you can save by incrementally using what you save to purchase more components for your solar system. For instance, if you were to rewind this about a decade ago you would find that the solar panels were significantly more expensive than they are now considering you could purchase a solar panel that provided 200 Watts for nearly $1,000 per panel. However, if you were to look at the same 200 Watts today, you would likely be able to get that for anywhere between $100 to $300. This shows a significant decrease in the cost to purchase solar panels as the technology improves.

On the other hand, you then have your mostly monopolistic

electrical system. I know most people don't think about their electricity

provider besides how much they have to pay but in the United States of

America, the government places a limit on how much they can increase

their rates because people use electricity all the time. Therefore, the

problem then becomes an issue about how much is the United States of

America limiting the rate increase? Well, usually it's about 8%, and

electric companies jump 8% in price almost every year. The electric

companies state that this is the only way that they can continue to make

money. When you install a solar power system, you pay an upfront cost,

and avoid all future rate hikes from your power company. For example,

if your company is now charging you around $0.11 per kilowatt of

power and you expect an 8% increase over the next decade, 8% * 10 is

actually 80%, which would mean that the rate would then go to about

$0.198 per kilowatt. However, the math is never that simple. Instead,

it's compounded annually by 8%. Therefore, we do this:

Year 1: 0.11*0.08=0.0088 + 0.11 = 0.1188

Year 2: 0.1188*0.08=0.009504 + 0.1188 = 0.128304

Year 3: 0.128304*0.08=0.01026432 + 0.128304 = 0.13856832

Year 4: 0.13856832*0.08=0.0110854656 + 0.13856832 = 0.1496537856

2	Year 1	0.0088	0.1188
3	Year 2	0.009504	0.128304
4	Year 3	0.010264	0.138568
5	Year 4	0.011085	0.149654
6	Year 5	0.011972	0.161626
7	Year 6	0.01293	0.174556
8	Year 7	0.013964	0.188521
9	Year 8	0.015082	0.203602
10	Year 9	0.016288	0.219891
11	Year 10	0.017591	0.237482
12			

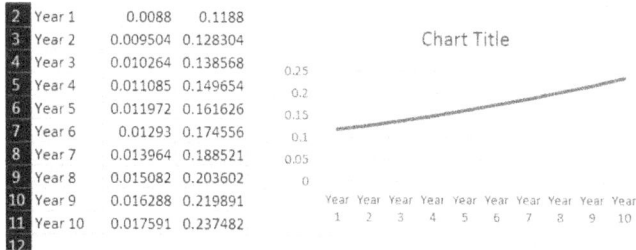

Chart Title

As you can see, this represents a significant change in pricing and also represents a huge hit to your wallet. By providing your own power, the amount of money that you save on electricity will increase over the years. Therefore, when you calculate how much you are saving by having a solar power system, you have to consider how much you would save over the lifetime of the system including any future rate hikes.

WHERE SHOULD I SET UP MY SYSTEM?

Panels Should Be in Sunny Areas

While this may be obvious, not everybody knows where the sunniest part of their house or yard is. Many people just assume that if they stick it out in the middle of their lawn, the sun will eventually hit it and power will be restored to their batteries. This is not a good idea because it takes time to charge batteries and so you need to find the sunniest part of your property to get the most out of your system.

If you have access to a large area to set up your panels and can afford it, the preferred configuration is to have one set of panels facing one direction and another set of panels facing another direction. This configuration will allow you to capture the greatest amount of sunlight throughout the day. However, this doesn't apply to most of our properties, and so it is also important to figure out which side of our property has the most sun on it during the day. Depending on where this is on your property you can then decide to either make stationary solar

17

panels that will absorb the most amount of sunlight or you can set up peripheral solar panels that will bend towards the sun more efficiently.

To determine the best location you could just stand there all day and see which side of your house gets the most sun, but you can also just look on the internet to see when the sun rises and sets in your location. If the sun rises on the left side and it can hit your solar panels without anything getting in the way and lowers on the right side where something blocks the sun, then you want most of your solar panels to be on the left-hand side. This will give you the most amount of solar power that you can gain. If you have a property where the sun rises at the back of your house, then you likely want to have the solar panels at the back of your house but at an angle where each solar panel can receive solar power within a few minutes of each other, like this:

Keep Wires Easy To Access

As I already mentioned before, this type of system is easy to maintain and fix because you have a very small network of wires that you must troubleshoot. The problem is that many people like to hide their wires. They dislike seeing wires outside. The truth of the matter is that you always want to keep these wires where you can easily access them. You do not want to hide them in a manner that makes it difficult to access. You will most likely have to do some troubleshooting. Your wires should be available so that you can quickly locate and fix any issues. You want to make sure that all of your systems are working before you begin to tuck away the wires. You also want to make sure that the wires are still easily accessible. There will be times when a wire receives too much electricity at once and begins to burn, in which case you want to cut off that section of the wire and stop it from burning. Wire burns in a solar system are easy to take care of if you can access the wires immediately. Otherwise, you risk compromising the entire system. Until you know exactly how much wire you need to run through your system to deliver the power to all of your devices and the proper safety techniques utilized to keep a house safe, you should not be hiding the wires. It is a fire hazard, they can be difficult to access in an emergency, and it turns an easily manageable system into something that is a nightmare to maintain.

Keep Monitoring Panels in Easy To See Areas

Just as you want to keep the wires in an easy to access place, you also want to keep the monitors in an easy to access place because they will be telling you exactly what's going on with your system. Depending on how you set up your monitor, you will be able to see the electricity coming in and the electricity going out. Usually, you have an under-charge controller that protects the batteries from being depleted of its entire energy reserve so that the batteries don't become fully dead. This can usually lead to the degradation and eventual death of the battery. The controller is located between the wires going out of the batteries and into the extension cords that allow you to plug in multiple devices. On the other hand, you also have an over-charge controller on the side where the solar panels connect to the batteries. You don't want to overcharge the batteries, especially lithium-ion batteries, because bad things happen when battery sources have more energy than they can store, such as becoming unstable and exploding. These two different charge controllers usually represent the main monitors that you have, but you can also have a secondary monitor that shows you what the controllers say from a different position in your house. You can now keep an eye on what's happening to your battery. If your solar panels are not providing enough electricity to your batteries, then your monitor will flash a warning so you can begin to restrict what devices are being used so you have enough electricity until you fix your problem, or wait until the next day when the problem is not as bad. You will not have 100% of your electricity available to you at all times due to local weather conditions, or your solar panels could accidentally fall out of place, or any other number of things. Therefore, it's important to keep an eye on your electricity so that you never run your batteries too low, even though you're never going to be able to use 100% of your stored electricity.

Batteries Should Be In Low Heat

Most batteries will have a warning label telling you not to house them in an environment that is very hot because batteries become much more active when heated and could eventually explode depending on the type of battery. What they do not tell you is that the colder the battery is, the less the chemical reaction it needs to provide you with a charge. Keeping it in a cool area will increase its life and increase the charge input and output of your batteries.

When you have something inside of a cold environment, the atomic structure of that item moves slower because of how cold it is. Your batteries will always lose electricity and generate heat due to the resistance of the wires and the battery itself. Cold helps by enveloping this heat and keeping the temperature down so the heat generated by the resistance cannot build up and damage the batteries. Remember, the hotter the environment is, the more active the electrons become, and the easier it is to lose electrical power and generate heat to that environment when it meets resistance.

Batteries Should Be In Low-Risk Conditions

We've already covered that batteries should not be in heated areas, but they should also be protected from other hazards within their environment.

A perfect example is the average rebuildable e-cigarette, which utilizes a battery similar to those used in solar batteries known as the 18650. There were a few stories where the batteries inside of self-made e-cigarettes have exploded on the users. In a lot of the cases, it was because the battery wrap was ripped and the battery itself was exposed. The average person deals with AA batteries, AAA batteries, and all forms of "normal" batteries that actually have a small motherboard in them to control the flow of electricity leaving the battery. In fact, most of these smaller batteries still have the battery wraps on them but are significantly safer because they have a motherboard controlling the charge coming out of them. In these e-cigarette batteries, the charge is not controlled, and this allows the batteries to disperse a lot of energy at a given moment making them useful in e-cigarettes which need variable voltages and amperages. However, because these lithium-ion batteries do not need a motherboard, they also come with the risk of becoming too overcharged on one of the ends of the battery. It has been shown that if you have an unwrapped lithium-ion battery such as the 18650 and if you poke it with a metal rod on the smaller end of the battery then the battery will gain an overcharge on that side and become unbalanced. Given some time, that battery will eventually explode on you. This is why you always want to make sure that the battery wrappings and cases are properly secured and away from sharp or magnetic objects.

BATTERY SELECTION

Lead Batteries

The benefits of most lead batteries lie in the cheapness and the versatility of the battery. Most lead batteries are cheaper than their lithium-ion counterparts, but it does come with some drawbacks, such as the suggestion that you never go past 80% of what your battery is capable of holding. This is because the battery will degrade over time as a result of being maximized and then fully discharged. The more you cycle through the recharge states, the more the battery will be unable to sustain its original maximum capacity. However, it is usually significantly cheaper to purchase one of these lead batteries for about $400 to $500 when the lithium-ion counterpart that would match the capacity of that lead battery is generally in the ballpark of anywhere from $800 to $1,000 right now. As technology progresses, the cost benefit of the lead battery actually might go down due to the recent work being done by Tesla making the massively improved versions of

25

their lithium-ion batteries significantly cheaper to produce and their ability to provide higher rates of cycling.

Needless to say, lead batteries are the most commonly used in all of the different types of solar batteries out there on the market. The best part about this is that it's pretty easy to dispose of these toxic batteries. Because the batteries that you are using in your solar power system are similar to those that are used in cars, so you can take them to automotive stores or specialty recycling centers and simply pay a fee for them to dispose of it for you. These fees pay for the cost that it takes to clean everything and recycle most of the parts so that new batteries can be made from the old ones, but that's a different discussion for a different time.

Lithium-ion Batteries

In a situation that is almost comical, lithium-ion batteries provide the same dilemma that you have if you were to buy rechargeable batteries versus disposable batteries. Rechargeable batteries are usually significantly more expensive than the disposable batteries, but because they can charge up to 1000 times more than disposable batteries. This is almost the same dilemma that you have whenever you're buying batteries for your solar system. Most of the time you have access to both the lead batteries and lithium-ion batteries. While lithium-ion batteries are the more expensive battery, they are capable of being recharged more than their lead counterparts. However, they also have complications at their disposal because it depends on how they were made. If they are a lithium-ion battery is primarily composed of organic components, then the battery is free from toxins and can just be thrown in the trash can. On the other hand, if the battery is made up of inorganic material then the odds of it being toxic are greatly increased. In most cases, recycling simply isn't an option because few recycling plants accept them as they are not recyclable components. This is unlike lead batteries, which can have the electrolytes cleansed or the packaging cleansed to be used in new batteries that are sold on the market.

Flow Batteries

Flow batteries are pretty new and hold some really great promise but the complexity of the reflow battery or redox battery or flow battery, whichever one you want to call it, makes it very difficult to use in any small installation. This is because these types of batteries currently need things like pumps, control units, and even containment vessels just in case something bad happens. However, this is a non-toxic type of power source and its cycling capacity (the amount in which you can recharge it) is currently not measurable and some even say that it has the potential to be rechargeable forever. Even better, you get the benefits of the lithium-ion battery, which you can recharge all the way to its maximum and then deplete it all the way to its minimum. Many people are looking forward to the introduction of flow batteries into the common market, but right now they are only available for enormous enterprises that can handle the hazardous conditions that they require.

Nickel-Cadmium Batteries

Nickel-cadmium seems like a very good way to go for most people because it has an excessively long life-span, of typically a decade or more. They are inexpensive to develop, and some of the technologies developed around nickel-cadmium allow you to essentially buy one and never have to worry about doing maintenance on them. The last part becomes the sole focus of whether you want to invest in this or not because nickel-cadmium is very toxic and hazardous, and many governments have limited the number of places that you can utilize them. Unlike the previous batteries which you can take for recycling, for nickel-cadmium batteries you have to remove the toxic material before you can properly dispose of the battery. This means that nickel-cadmium is very hard to buy in large quantities and you may not want to risk yourself or your family just because you want a cheaper more quality battery than the safer types of batteries. Nickel-cadmium has actually been around for about a century, but because of how toxic it is and because it is a lot less powerful than the other batteries that we've mentioned, it's not very widely used.

Battery Crates

The final thing to think about when you are considering your batteries is the type of storage that you need to buy in order to house them. As I mentioned before, storing your batteries in a colder environment is actually better for you in the long run because it provides the batteries with the optimal temperature in which to work. Additionally, having batteries at a hotter temperature makes them more volatile and more likely to explode on you. Lastly, you want to make sure that whatever you have in terms of containing them has no pointy spots because anything that could rupture the container of the battery could potentially set the battery on an unstable course of charge control.

Therefore, there are three common types of housing solutions when it comes to batteries. The first one is to simply house them inside of a metal container and put them in a cold environment such as a basement or an air-conditioned garage. The metal container makes sure that nothing will ever come in contact with the batteries and you can usually install metal containers on the sides of walls where they won't be in the way of anything that you're trying to do. However, making such a container can usually be fairly expensive depending on how many batteries that it needs to hold and it can be unwieldy whenever you're trying to take batteries out and put new batteries into the system. Therefore, metal containers are usually custom designed and built for the person who needs them.

The most common solution is to do everything with wood because wood generally insulates your batteries from any heat outside of the crate without making it too hot on the inside. Also, wood generally only splinters whenever it's moved around a lot, and you shouldn't be moving these around anyway. This means that you can install it on wooden walls, and you can generally have it wherever you want it. You need to consider how heavy the batteries are so that the wood doesn't break.

The last type of housing storage, which almost no one who is experienced in this technology does, is to just have these batteries out in the open. This leaves them vulnerable to puncturing, interacting with magnetic objects, or anything else that could potentially go wrong. This is usually seen in more inexperienced individuals because those individuals have had to only deal with disposable batteries that you can leave out in the open without having to worry about what's going on with them. A variant of this type of loose solution is actually to have your batteries out in the open but on movable trays. They are still open to puncturing and interacting with magnetic objects, but the option to move them allows them to be stored in areas that are predominantly not used inside of the house so that those chances are less likely. Additionally, it does make it very easy to work with them and if you don't have anything sharp around then it's actually a pretty good idea. However, it is preferable to have them inside of a wooden container so that you can section them off and if they do become unstable, you do have that wood to take in the initial blast if they so happen to explode.

A Short message from the Author

Hey, are you enjoying the book? I'd love to hear your thoughts!

Many readers do not know how hard reviews are to come by, and how much they help an author.

I would be incredibly thankful if you could take just 60 seconds to write a brief review on Amazon, even if it's just a few sentences!

Please head to the product page, and leave a review as shown below.

Thank you for taking the time to share your thoughts!

Your review will genuinely make a difference for me and help gain exposure for my work.

PV SELECTION

What does PV or Photovoltaic mean?

I really like how industries try to certain words sound special. What photovoltaic means is it is a solar power system. However, there is a selection of photovoltaic systems out on the market and which one you choose depends on what you want to do when it comes to electricity. In this section, we'll go over the different photovoltaic selections so that you can pick one that is right for you.

Non-hybrid Integrated

The first selection is the non-hybrid integration method, and this means that you want to get off the electric grid entirely. There are some downsides to choosing this type of option. Some areas don't let you sustain an independent electric grid and many areas are currently in legal battles because they want to have this option. Non-hybrid means that you do not share your electricity with anybody else and you are no longer connected to the massive electric grid that your electric company provides. A lot of people who live in the desert or in the woods that don't have immediate access to the electric grid, and don't want to pay the connection costs for their property, prefer this non-hybrid integrated system. On the other hand, many people who don't want to deal with the electrical business anymore try to do this in cities where it's illegal to not be on the electrical grid. The way that the city usually finds out that a house is not on the electric grid is if the local power company reports that this person is no longer on their grid. The reason they want you to be on the grid is that the electricity provider takes the proper safety measures and has the proper certifications to prove that they know what they are doing, whereas the average individual DIY solar person does not.

Hybrid Integration

Hybrid integration is where you get the approval of the electrical company that you are paying to build your system into your house but also feed electricity back into the grid. Essentially, you build the same solar system layout, but this time you are allowed to feed electricity back into the grid. Hybrid integration is usually carried out by a company because of the necessary needs for certification and quality assurance and insurance. The electric company doesn't know whether you are properly qualified to feed electricity back into the system without causing complications. Companies become certified in this so that they can provide the needed expertise to get the electrical company to approve of this.

Modular

The modular system is the most common type of solar power system on the market because it's the easiest to do by hobbyists and requires no certification to do it. It allows people who want to build the system themselves to create the system and put most of their electronics on the system without needing to gain the approval of the electric company. This means that you can take the power that's needed for all of your devices and put it on a grid that you built but you would still have the monthly charge of having a meter on your house for the electric company. Needless to say, you would still have to pay for any lights that are used inside of the house since those still run on the electrical grid owned by the electric company. However, most people make a modular system that doesn't require the use of these lighting

systems in their house and so they generally just have to pay for renting

the meter unit that's on the side of their house.

A Word of Warning

If you decide to make a non-hybrid integration without the

proper certifications, you will make your house unsellable unless it is

inspected by an electrician that has the certification for solar grid

systems. Electric companies love to corner you with required

certifications and approvals so that it is difficult to get off the electric

grid. If you decide to put an electric grid into your house you run into a

wall of certifications and approvals but, depending on how you do this,

you could also run into damages. You see, the electric company has

tried to pull a type of charge where if you make the system integrated

on your house and you had to modify the line going through the meter

to your house, then the electric company charges you for damages to

their meter. As unjustified as this is, they have gotten away with it quite

a few times.

INVERTER SELECTION

What is the Inverter?

An inverter is a device that takes in a bunch of wires from a power source that is usually a digital current and changes it to an alternating current, which is what is utilized by your household and your devices. These usually go by the amount of hours that can be channeled through the inverter and this is simply calculated by multiplying the amount of voltage that you have by the amount of amperage that you have that's going through the system at the time. Almost all inverters do the exact same thing unless you're talking about an alternating current inverter, which is almost never applicable in solar power installations because almost all devices and home wiring is based on the alternating current

The Three Kinds of Inverters

Sine Wave

This is the most preferred type because it allows a very fine control over what you need and it generally is provided by the local electric company. In other words, this is the type of inverter that you actually want in most cases and buying anything less than this means that you are limiting the number of items that can go on that inverter. It's called a sine wave is because this is what you see you whenever you hook it up to a device called an oscillator.

Modified Sine Wave

A somewhat square dip represents a modified sine wave up and below with a slight step in between. This is the middle point between the regular sine wave and the square wave. It actually works with a lot of devices but since it isn't as refined as the sine wave, you will use a lot more electricity than you would have if you would have bought a regular sine wave inverter. Additionally, modified sine waves are not good for electronics that are sensitive, such as phones or computers.

The benefit is that this is cheaper.

Square Wave

This last type of inverter is so uncommon that most people don't even

have it but essentially the wave is square, and it is really only good for

basic motors. These are the cheapest you will find.

OTHER CONSIDERATIONS

Charge Controller

While they are not mandatory for a system to be built and used, charge controllers are an excellent option for protecting your system. There are two different types of charge controllers: an under-charge controller and an over-charge controller.

An under-charge controller protects your battery from being drained past the point of recovery because lead batteries are known to only be useful for 80% of the batteries total capacity. If you go past 80% on a lead battery, this will lead to a severe degradation of the battery where you are able to use less and less of the battery until you eventually kill the battery. Additionally, the lithium-ion batteries that I spoke of earlier can become unstable when undercharged by simply drawing too much of the electricity, lowering the voltage. You see, as you withdraw electricity from a lithium-ion battery in devices like your cell phone or your e-cigarette, they will lose voltage over time, and this means that if their voltage drops below a certain amount, these batteries are under-charged or are considered dead. This is why an under-charge controller is very useful.

On the opposite side of the spectrum, you have an over-charge controller which helps protect you from things like lightning storms where lightning can hit your solar panel and supercharge the voltage going through the wires. Needless to say, this is going to make your wires erupt in flames but if you have an overcharge controller then the likelihood that the massive increase in voltage hits your battery significantly decreases. If lightning strikes a solar panel, you will likely have an increase in voltage amongst the solar panels that are beside the ones that were hit. Therefore, you will likely have multiple panels affected by the lightning strike and all of the wires will be sending a massive amount of electricity through to the system. Additionally, if you've connected those through a series, then all of those that are in that series will be affected as well. The over-charge controller helps to prevent any electricity from going into the system if it is over a specified amount on the charge controller. This is different from protection, which actually doesn't control the level of charge coming through. However, because it is controlling the amount of charge coming through that connection, it effectively protects your batteries to a certain degree.

Hand-built vs. Factory Built

There isn't really a fight between those who build a solar panel system themselves and those who purchase it from a factory because the people who build it themselves are also usually installing the system themselves. Those who are buying from the factory usually have somebody else install the system for them, and this means that they are going to call the person that installed it to troubleshoot it rather than trying to troubleshoot it themselves. That doesn't mean that all of those who buy factory-built solar panels will be like this, just the majority of them because it's easier.

However, what is the benefit of building it yourself over having someone else build it for you? Well, the problem with someone building it other than yourself is that you cannot control the design of the actual solar panel itself. Your solar system is dependent on how those solar panels are designed. For instance, if you decide to build it yourself you can actually buy the solar cells for very cheap and you can connect those panels in whatever parallel or series that you want. On the other hand, you just know that this factory-built solar panel will provide you 1000 Watts at about 8 amps. It's mathematically easy to figure out the voltage in this case, but the truth is that if you need to supply power to a 12-amp battery, then you are going to need to find another factory-built solar panel that will compensate for the missing 4 amps. As you can see, this can get quite cumbersome, and this is because you do not have direct control over how these solar panels are built. There's no way to special order your solar panel grids. Instead, you have to deal with what's available and compensate for what you're missing.

Additionally, they may not even allow you to set up serial or parallel panels because of the connections that they use. Some of the factory-built solar panels don't even allow you to connect to other solar panels and just have a socket on the back that allows you to put in a standard plug. The idea behind this is quite ridiculous, but someone who doesn't understand how to build a solar power system would simply think that this was the normal way that solar panels were built. Lastly, factories attempt to make the cheapest possible panel they can that is within the safety parameters to maximize their profit. That means that you are likely not getting the best, most optimized version of a solar panel. By building it yourself, you can select the solar cells and the configuration that is used in your solar panel.

Output Readers

Output readers are actually almost as useful as the monitors themselves. You see, the monitors provide you with a view of your entire grid system and normally the bare minimum monitoring is simply to have an under-charge controller and an over-charge controller. However, if you have panels that are hooked up in serial with each other and you have panels that are hooked up in parallel with each other, output readers are very handy in telling you which panels are experiencing problems. How an output reader works is it takes in the wires from one side and feeds electricity through a motherboard, utilizing a very small percentage of the power, before sending it out through the other side of the output reader. By feeding the electricity through the motherboard you can actually read how much voltage, amperage, and, ultimately, the wattage is coming from a single solar panel. Therefore, if you are finding that your solar power system is having a problem, then you can simply glance at each of the output readers and see if it is a system-wide problem or if a few of the solar panels are causing the problem. It provides an extremely easy way to find issues, and output readers are usually very cheap. The specific Google search words you need to use are *LED Voltage AMP Gauge Display*.

Monitoring

A good example of a monitoring solution is the MOXA W321. A solar power system monitor is very similar to an output reader because the primary purpose of the monitor is to observe how much electricity is going out of the system itself. Depending on your monitoring solution, you may be able to monitor the output at certain levels in the day so that you optimize how much electricity you use to what you receive. Likewise, many of them tell you exactly how much power you are a pushing out rather than just how much voltage you are

putting out or how much amperage you are putting out. The purpose of

the monitoring solution is to give you a quick overview of everything

that's happening inside of your power system so that you don't have to

go to every output reader individually and see exactly how much each

one is providing. Output readers are really good at determining which

solar panels are causing problems and which ones are not. Depending

on the sophistication of your monitoring solution you can actually

compare the amount of power that you generated today to the day

before, along with a weather record during the same period. This would

be helpful in a situation where you are trying to determine why you

generated less power one day to the next. Your monitoring system will

show you that the cause of the reduced electricity was the weather, so

you do not need to check your panels for damage.

You might think that you wouldn't need something like this but

how often do you actually keep track of what the weather was like week

by week? As I already said, it really does depend on the monitoring

solution that you've chosen because some just tell you what the voltage, wattage, amperage, and watt-hours your system is providing, and others keep records of how much you provided a year ago versus how much is being provided now. The more useful monitoring systems are often the more complex ones because they allow you to optimize and keep track of what's going bad and what's staying good. Additionally, it can actually tell you when one of your inverters have failed or whenever one of your charge controllers is not providing the level of control that it's supposed to before it's too late.

Charge Protection

I'm pretty sure that you've gotten the gist of why all of these safety mechanisms are in place, but you have one final option that you can use as a protection and a way of making sure that you have a little extra stored for emergencies. This last step of protection is to get a multi-socketed surge protector extension cord, and we use these in our daily lives. In fact, around my house I believe I have somewhere around 7 of these without being on a solar power system, and it's because two sockets are usually never enough for the number of electronics that I run. However, in a solar power system you can have these as the last form of protection because most of these have surge protectors built into them that are pretty beefy. In fact, you can go with a solution that's even better than just having multi socketed surge protector extension cords, and that is to have an alternate power supply (APS) which is often also referred to as a UPS. Most of the individuals that utilize these types of power supplies are people who are on their computer a lot and want the opportunity to close and save programs when the power goes out so that they are not badly affected, but their purpose works very well for us. Due to the fact that it stores additional power, you can use this as both a form of protection and an alternate way of getting power out of your solar power system. Once you are providing a solid amount of power then this system does not kick on, usually, until the power has been turned off from the system. Then, everything connected to the system now gets power from the alternate power supply. As I said, it's very useful if you upgrade to this but it's not necessary.

STEP BY STEP SETUP

Buying a Panel

If you're buying a panel from the factory, then the first thing that you want to consider is if you want to buy one that allows you to have access to the black and red wires. Having access to the black and red wires allows you to build up the solar panels in a serial or a parallel connection. A serial connection allows you build up the voltage in a system, but parallel allows you to build up the amps in a system. To do most of your calculations, you will need this: Watts = Volts * Amps.

Gauging Your System Needs

The second step that you need to do is actually gauge what your needs are. You might think that this is simply calculating all the wattage that you need inside your house, but this doesn't really matter in the context of voltage and amperage. You need to have a system that matches the amperage of the devices that are being put on that system. The beauty about extension cords and most devices is that most of the devices generally follow the same amount of amperage, and most extension cords will actually set the amount of amperage that's coming through, but you shouldn't rely on this. Instead, you should calculate all the devices that need the same amount of amperage and then separate them into grids and calculate the wattage needed by those devices. Each of those grids will represent different segments of battery supplies that you need in order to connect those devices to a proper grid of electricity. It is important to understand that you can also buy over-charge controllers and under-charge controllers that will change the level of the amperage so that you don't have to make as many grids, but the ideal situation if you don't have such controllers is to separate them by the amount of amperage that they need.

Battery Supply, Inverter and Panel Setup

Once you have found all the needs of your system, you need to set up your batteries so that they can supply the amount that those systems demand. It's important to realize that if your system needs more voltage then you need to connect them in serial, and if your system needs more amperage then you need to connect them in parallel. A serial connection is when you connect the positive to the negative or the negative to the positive over and over. So, if you have three batteries that have 6 volts each, and you connect each battery's negative to the next positive until you reach the third battery then the first battery will channel all the different voltages and this would make 18 volts.

A parallel connection, on the other hand, is when you connect positives to positives and negatives to negatives. This is utilized in order to increase the amperage that you were trying to gain in a system. The important part is that when you are gauging your battery set up, you need to make sure that your amperage and your voltage match the devices, which is why it is very useful to have a charge controller.

On the same spectrum, once you have your battery supply set up you now need to set up your panels in much the same way. You need to make sure that your panels are capable of providing the power necessary to your battery supply grid so that it can provide the electricity that your electrical grid needs. This means that you'll need to connect some of the panels in parallel and some of the panels in series in order to match the needed requirements.

Now comes the inverter because it depends on your preference as to where you want to place the inverter. Some people prefer to place their inverter before it gets to the batteries so that you can change direct current electricity into alternate current electricity to feed it directly into the wiring in the walls. With the Battery Solution, you can use the inverter to switch from the battery banks direct current to the alternate current that is required by devices and by your entire house for that matter. Remember, the point of the inverter is to turn the solar power's direct current or the battery power's direct current into an alternating current so that it can be used by the devices and the house whether it is connected to the wiring in the house or it is connected to the devices themselves.

Under-charge and Over-charge Controller

The next step in this process is to connect your under-charge controllers in between the batteries and the places where you are providing the power output, in most cases, this would be the extension cords that have multiple sockets. This means that if you have this type of under-charge controller, which means that it is not built directly into the wiring of the house, then you will need to have an under-charge controller that is capable of taking in plugged in sockets. Then you have to switch over to the over-charge controller, which should be connected between the solar panels and the battery supply grids. You don't need to make sure that all of the solar panels are connected to the over-charge controller, just the wires that are going into the batteries themselves.

Connecting to a Monitoring Station

At this point, the system is actually complete and you should be able to provide power, but if you want to provide a monitoring solution then this is the time where you intercept the connection that would normally go in between the under-charge controller and the batteries so that you can insert it. Normally, you would have one enormous collection of wires connecting into a single monitoring solution unless you have connected them to an inverter.

CAN I MAKE IT PORTABLE?

It Depends on Size

The first part about whether you can make your system portable or not is actually the size of your needs versus the size of the system that you want to build. If you are looking to build a grid that can feed 1 kilowatt of power, then the odds are high in making it portable because you just need a single panel that is capable of producing one-kilowatt of power that can be fed into a battery that can hold it. However, if you are like the average household that utilizes somewhere around 3kwh in terms of power, then it becomes quite unwieldy to make this portable so that you can take it on an RV or something. Additionally, when we talk about portable, we are talking about a solar power system that can be installed on a RV, or a smaller mobile system that can be set up to provide power at various locations. No solar power system is currently strong enough to be used as a completely portable solution unless it is tied to a stationary object large enough to house at least a single panel.

It Depends on Efficiency

You also have to take into account just how efficient your solar panels will be because solar panels that were developed a decade ago are nowhere near as efficient as the solar panels that are being produced today. In fact, most of the solar cells that you will be buying on the market, if you decide to build your solar panel, are actually solar cells that were taken from panels built years before. The design of these solar cells has changed over the years, and this is how you can tell the different grades and when they were made. The most common type that you will find on the market are the rectangular ones, as this was the very first type of commonly sold solar panel cell. These are not very efficient and while they are very cheap, they don't convert power very well in comparison to the ones that are more octagonal, because the octagonal ones are actually newer and are capable of a higher efficiency rate. The efficiency matters because the less efficient your solar cells are, the more of those solar cells you will need in order to meet your electrical needs.

It Depends on Where You are Going

The last part of this equation is actually about where you are going because some places simply aren't good for those who want to depend on solar panels. For instance, one of the worst places you could go is Alaska because Alaska is cloudy almost year-round and the amount of sun that you actually get in Alaska is roughly half or less than you would get in some place like Florida. This is important because if you get less sunlight, then you will need more solar panels in order to compensate. The problem is that you eventually get to the point where you need so many solar panels that it becomes pointless. It becomes beneficial to just go to an RV park and plug into the electrical socket that they provide and simply draw from the main grid rather than your solar system.

CONCLUSION

Welcome to the end of this book. It's important to note that this is not everything that there is to know about building a solar power system. We could have covered how to make your own solar power system. As new technologies come out and as we find different ways of better managing our power supplies, this field will continue to change for better or worse. However, I want to thank you for reading this book, and I look forward to helping you next time.

The end… almost!

Reviews are not easy to come by.

As an independent author with a tiny marketing budget, I rely on readers, like you, to leave a short review on Amazon.

Even if it's just a sentence or two!

So if you enjoyed the book, please head to the product page, and leave a review as shown below.

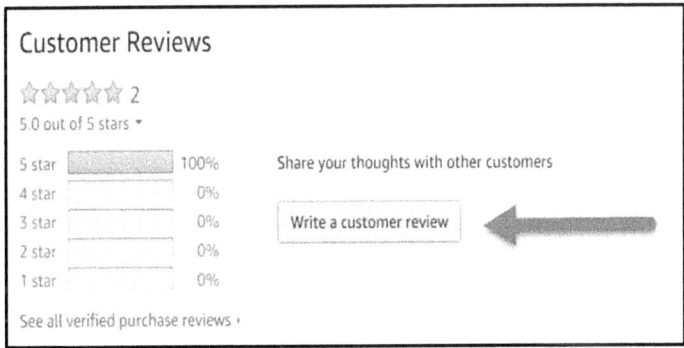

I am very appreciative for your review as it truly makes a difference.

Thank you from the bottom of my heart for purchasing this book and reading it to the end.

www.ingramcontent.com/pod-product-compliance
Lightning Source LLC
Chambersburg PA
CBHW071516210326
41597CB00018B/2778